A Guide to MEDICINAL PLANTS of Appalachia

FOREWORD

THE MEDICINAL or therapeutic uses of the plants described in this guide are not to be construed in any way as a recommendation by the authors or the U.S. Department of Agriculture. Some of the dried crude drugs, which must be modified considerably before commercial use, can be extremely poisonous when not used properly. Readers are cautioned against using these plant drugs for purposes of self-medication.

Besides descriptions of 126 medicinal plants of the Appalachian region, this guide includes a glossary of the terms used, a reference list of publications, and a listing of additional source material.

CONTENTS

INTRODUCTION	1
PLANT DESCRIPTIONS	3
Identification	3
Names	3
COLLECTING PLANTS	4
Time of year	4
Areas	10
Tools	10
PROCESSING	11
Cleaning	11
Drying	11
Packaging and storing	14
COLLECTING POLLEN	15
Methods	15
Drying	16
Grass pollen	17
REFERENCES	18
GLOSSARY	21
GUIDE TO THE PLANTS	27
INDEX OF COMMON PLANT NAMES	281

INTRODUCTION

DESPITE INCREASES in the production of synthetic drugs, natural plant drug materials are still economically significant in the United States, and large quantities are harvested in the southern Appalachian region each year for medicinal purposes. A 1962 survey of 328,599,000 new prescriptions written in the U. S. showed that 25 percent were for drugs from natural plant products.

However, during the past 30 to 50 years, fewer and fewer people have been harvesting wild plants in Appalachia, which is the principal American source, mainly because of families emigrating to more prosperous areas. Between 1950 and 1960, the southern Appalachian region lost through emigration more than a million people, nearly a fifth of the population. Increases in local blue-collar employment opportunities, a growing reluctance to work in the fields and forests, scarcity of some plants because of over-collecting, and land-use changes have also reduced the natural plant harvests for drug materials.

To locate, collect, and prepare plants for market is time-consuming work. Some collectors do not know all the useful plant species and the best markets for them. This manual was prepared to help collectors identify, collect, and handle plants, plant parts, and pollen.

Not all the plants listed are marketable at all times; so the collector would do well to write to one of the collecting houses listed (table 1) for prices and information about market demand. Buyers of such material are helpful in providing other useful information on collecting.

Table 1.—Names and addresses of buying houses*

Names	Addresses
PURCHASERS OF BOTANICALS	
Blue Ridge Drug Company	P. O. Box 234, West Jefferson, North Carolina 28694.
Coeburn Produce Company	Second and Grand Streets, Coeburn, Virginia 24230.
C. R. Graybeal	Roan Mountain, Tennessee 37687.
F. C. Taylor Fur Company	227 E. Market Street, Louisville, Kentucky 40202.
Greer & Greer	Box 307, Princeton, West Virginia 24740.
Greer Drug & Chemical Company	P. O. Box 800, Lenoir, North Carolina 28645.
Nature's Herb Company	281 Ellis Street, San Francisco, California 90025.
Old Fashioned Herb Company	581 N. Lake Avenue, San Francisco, California 90025.
Smoky Mountain Drug Company	935 Shelby Street, Box 2, Bristol, Tennessee 37620.
Wilcox Drug Company	P. O. Box 391, Boone, North Carolina 28607.
Wilcox Drug Company, Inc.	Box 470, Pikeville, Kentucky 41501.
IMPORTERS THAT BUY, SELL, AND PROCESS BOTANICALS	
Hathaway Allied Products	2024 Westgate Avenue, Los Angeles, California 90025.
S. B. Penick & Company	100 Church Street, New York, New York 10007.
VENDORS OF DRUG AND HERB SEED AND OTHER PROPAGATING MATERIALS	
Gardens of the Blue Ridge	Ashford, North Carolina 28603.
Harry E. Saier	Dimondale, Michigan 48821.
Indiana Botanic Gardens	P. O. Box 5, Hammond, Indiana 26325.

*These firms are mentioned for information only, and this mention should not be considered as an endorsement or recommendation by the U.S. Department of Agriculture or the Forest Service.

PLANT DESCRIPTIONS

Identification

To help the collector identify plants, brief descriptions are given in this guide. Some closely related plants, such as *Lobelia* (Indian tobacco), are difficult to identify before the seed capsules have developed; so as a further aid in identification, sketches or photos accompany every plant description.

A collector who wants to identify a plant known only by a common name should locate that name in the index and then refer back to the descriptions and illustrations to identify the plant. If the same common name is applied to more than one plant, this will be shown by the page numbers next to the common name in the index.

Names

Plant names can be confusing. A plant may have many common names, and the same common name may be applied to several unrelated plants. We have tried to show as many common names as possible, listing first the preferred common name suggested by the Subcommittee on Standardization of Common and Botanical Names of Weeds. If this list did not include names for a plant, we used *Standardized Plant Names*. Other references used were *Flora of West Virginia, Manual of Cultivated Plants, Flora of the Northeastern United States,* and State experiment station bulletins.

Scientific names are also given to simplify proper identification of plants. Although a number of common names may be in use for a given plant, only one scientific name is used.

COLLECTING PLANTS

Time of Year

It is important to collect at the time of the year when the drug contents of the plants are at their peak.

Roots are collected either very early in the spring before growth has begun, or late in the fall. Herbs (the part of the plant above ground) are usually collected during the blooming-fruiting period. Leaves are usually collected before blooming begins and can either be removed from the plant in the field, or the plants can be harvested and the leaves can be removed later at a collection area. Seeds and fruits are best harvested when ripe. Bark should be collected when it slips most easily, during the dormant season or in early spring.

The parts of each plant collected are shown in table 2.

Table 2.—The parts of plants collected

B	=	Bark.	FL	=	Flowering top.	RR	=	Roots, rhizome.
BR	=	Bark of root.	H	=	Herb.	SE	=	Seeds.
BU	=	Buds.	J/S	=	Juice or sap.	ST	=	Stalk.
EP	=	Entire plant.	L	=	Leaves.	T	=	Twigs.
F	=	Fruit.						

Plant	Part collected												
	B	BR	BU	EP	F	FL	H	J/S	L	RR	SE	ST	T
1. Acer spicatum Lam.	x												
2. Achillea millefolium L.							x						
3. Acorus calamus L.										x			
4. Adiantum capillus-veneris L.									x	x			
5. Adiantum pedatum L.							x						
6. Aesculus hippocastanum L.	x				x								
7. Aletris farinosa L.										x			
8. Alnus serrulata (Ait.) Willd.	x												
9. Amaranthus hybridus L.							x						
10. Angelica atropurpurea L.							x			x			
11. Aplectrum hyemale (Muhl.) Torr.										x			
12. Apocynum androsaemifolium L.										x			
13. Apocynum cannabinum L.										x			
14. Aralia nudicaulis L.										x			
15. Aralia racemosa L.										x			
16. Arctium lappa L.										x	x		
17. Arctium minus (Hill) Bernh.										x	x		
18. Arisaema triphyllum (L.) Schott										x			
19. Aristolochia serpentaria L.										x			
20. Asarum canadense L.									x	x			
21. Asclepias syriaca L.									x	x			
22. Asclepias tuberosa L.										x			
23. Baptisia tinctoria (L.) R. Br.		x					x			x			

CONTINUED

Table 2—Continued

Plant	B	BR	BU	EP	F	FL	H	J/S	L	RR	SE	ST	T
24. Berberis vulgaris L.	x												
25. Betula lenta L.		x											
26. Caulophyllum thalictroides (L.)										x			
27. Ceanothus americanus L.		x								x			
28. Chamaelirium luteum (L) Gray									x	x			
29. Chelone glabra L.							x						
30. Chenopodium ambrosioides L.				x	x								
31. Chimaphila maculata (L.) Pursh.				x					x		x		
32. Chimaphila umbellata (L.) Bart.				x					x				
33. Chionanthus virginicus L.	x	x											
34. Cimicifuga americana Michx.										x			
35. Cimicifuga racemosa (L.) Nutt.										x			
36. Cnicus benedictus L.							x				x		
37. Collinsonia canadensis L.							x						
38. Comptonia peregrina (L.) Coult.									x	x			
39. Corallorhiza spp.										x			
40. Cypripedium calceolus L.										x			
41. Datura stramonium L.									x			x	
42. Dioscorea villosa L.										x			
43. Echinacea purpurea (L.) Moench.										x			
44. Eryngium aquaticum L.										x			
45. Euonymus atropurpureus Jacq.	x	x											
46. Eupatorium perfoliatum L.						x	x		x				
47. Eupatorium purpureum L.						x			x				
48. Fragaria virginiana Duch.					x				x	x			
49. Fraxinus americana L.	x												
50. Galium aparine L.							x						

51. *Gaultheria procumbens* L.
52. *Gelsemium sempervirens* (L.) Ait.
53. *Gentiana villosa* L.
54. *Geranium maculatum* L.
55. *Hamamelis virginiana* L.
56. *Hedeoma pulegioides* (L.) Pers.
57. *Hepatica acutiloba* (D.C.)
58. *Hydrangea arborescens* L.
59. *Hydrastis canadensis* L.
60. *Jeffersonia diphylla* (L.) Pers.
61. *Juglans cinerea* L.
62. *Juglans nigra* L.
63. *Juniperus communis* L.
64. *Juniperus virginiana* L.
65. *Lactuca scariola* L.
66. *Leonurus cardiaca* L.
67. *Lindera benzoin* (L.) Blume
68. *Liquidambar styraciflua* L.
69. *Lobelia inflata* L.
70. *Lycopus virginicus* L.
71. *Marrubium vulgare* L.
72. *Menispermum canadense* L.
73. *Mentha piperita* L.
74. *Mentha spicata* L.
75. *Mitchella repens* L.
76. *Monarda didyma* L.
77. *Myrica cerifera* L.

CONTINUED

Table 2—Continued

Plant	B	BR	BU	EP	F	FL	H	J/S	L	RR	SE	ST	T
78. Nasturtium officinale R. Br.							x						
79. Nepeta cataria L.							x						
80. Panax quinquefolium L.										x			
81. Passiflora incarnata L.						x				x			
82. Phytolacca americana L.					x	x							
83. Pinus palustris Mill.								x					
84. Pinus strobus L.			x					x					
85. Plantago spp.									x				
86. Podophyllum peltatum L.										x	x		
87. Polygala senega L.										x			
88. Polygonatum biflorum (Walt.) Ell.										x			
89. Polygonum hydropiper L.							x						
90. Populus balsamifera L.	x												
91. Prunella vulgaris L.							x						
92. Prunus serotina Ehrh.	x												
93. Quercus alba L.	x								x	x			
94. Rhus glabra L.	x	x			x					x			
95. Rubus spp. L.	x	x			x					x			
96. Rumex crispus L.													
97. Salix alba L.	x		x										
98. Salix nigra (Marsh)	x		x										
99. Salvia officinalis L.							x						
100. Sanguinaria canadensis L.									x	x			
101. Sassafras albidum (Nutt) Nels.								x					
102. Scrophularia marilandica L.								x			x		
103. Scutellaria lateriflora L.										x	x		
104. Senecio aureus (L)							x						

#	Species											
105.	*Solanum carolinense* (L)			×								
106.	*Sorbus americana* Marsh.			×		×						
107.	*Spigelia marilandica* L.								×			
108.	*Stellaria media* (L) Cyrillo									×		
109.	*Stillingia sylvatica* L.											×
110.	*Tanacetum vulgare* L.					×		×				
111.	*Tephrosia virginiana* (L) Pers.						×		×		×	
112.	*Tiarella cordifolia* L.			×			×		×		×	
113.	*Trifolium pratense* L.							×		×		
114.	*Trilisa odoratissima* (Walt.) Cass.						×		×		×	
115.	*Trillium erectum* L.									×		
116.	*Tsuga canadensis* (L) Carr.		×		×							
117.	*Ulmus rubra* Muhl.		×									
118.	*Veratrum viride* (Ait.)					×		×	×		×	×
119.	*Verbascum thapsus* L.				×	×			×			×
120.	*Verbena hastata* L.							×				
121.	*Veronicastrum virginicum* (L) Farw.							×				
122.	*Viburnum nudum* (L).		×	×								
123.	*Viburnum prunifolium* L.		×	×								
124.	*Xanthorhiza simplicissima* (Marsh.)	×							×		×	
125.	*Xanthoxylum americanum* Mill.	×						×	×		×	
126.	*Xanthoxylum clava-herculis* L.	×					×					

Only four plants are listed from which sap is collected. *Lactuca scariola* provides a milky juice; the plants are collected in summer for extracting the juice from the stems. The other three are trees. *Liquidambar styraciflua* (sweetgum) exudes balsam into natural pockets between the bark and the wood. Excrescences on the bark are cut for collecting the sap. *Tsuga canadensis* (hemlock) produces an exudate of resin, which occurs on the stem in reddish brown, opaque, or translucent pieces. *Pinus palustris* (long-leaf pine) is a source of turpentine, pine oil, tar, pitch, and rosin.

Collectors are urged to leave enough plants growing in each locality to conserve the plant population for future years.

Areas

The most likely areas where each plant may be found are listed. Time and effort can be saved by narrowing areas of search to those habitats where the plant usually occurs. Residents of a region can often provide information about growing areas. Care should be taken to respect property rights of landowners, and permission should be obtained before entering private land. State and Federal laws regarding plant collecting should be checked for given localities.

Tools

A wide range of tools—from a pocket knife to shovels of one type or another (fig. 1)—can be used by the collector, depending on what plant parts are to be harvested. For example, a shovel or an asparagus knife would be needed to harvest roots—plus a pair of shears to cut the tops. Bark collectors need a sharp knife, the size depending on the thickness of the material to be harvested.

As the collector gains experience, he will be able to determine exactly what tools are required for each kind of material. However, he will always want to carry the minimum number needed. All tools should, of course, be kept oiled and sharp; and they should be cleaned after each use.

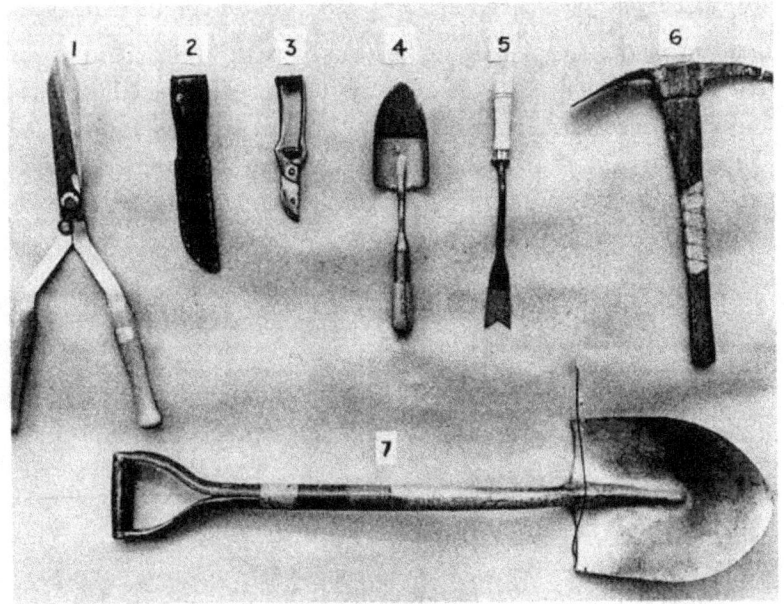

Figure 1.—Harvesting implements: (1) hedge clippers, (2) sheath knife, (3) pruning shears, (4) trowel, (5) asparagus digger, (6) pick, and (7) shovel.

PROCESSING

Cleaning

Cleaning harvested plant materials is called *garbling*. It includes removal of stones, soil, and unwanted plants and plant parts. Roots and underground parts may have to be washed if soil clings to them.

Drying

Rapid drying is needed to preserve green color, to reduce spoilage and molding, to reduce or stop enzyme action that destroys drug constituents of plants, and to make the materials more compact for shipping. This is important because improper drying can result in reduced value, if not complete loss of the material collected.

Two methods are used to dry drug plants. The natural method is the simpler; it makes use of natural air temperature and air movement. The second method uses artificially heated and circulated air. Many different types of equipment are used. These range from simple and inexpensive tools for handling small quantities of a few kinds of plants to the large costly equipment needed for handling many kinds of plants in large lots.

Natural drying.—Natural drying uses the sun's heat plus shade and air movement. A porch or barn floor or almost any shaded area with a dry floor will do. Shallow wire-bottomed trays are cheap to make and can be used to good advantage (fig. 2). Too much exposure to the sun can cause loss of green color, thus decreasing the value and marketability of certain materials.

Figure 2.—Shallow wire-bottomed trays can be used for natural drying.

Artificial drying.—A simple and inexpensive drying box can be built for less than $25, using a standard home-type, fan-driven space heater (fig. 3). This box takes up less floor space than the natural drying method, dries plants more quickly, and produces a more uniform product. Racks (fig. 4) provide space for drying all types of plant materials.

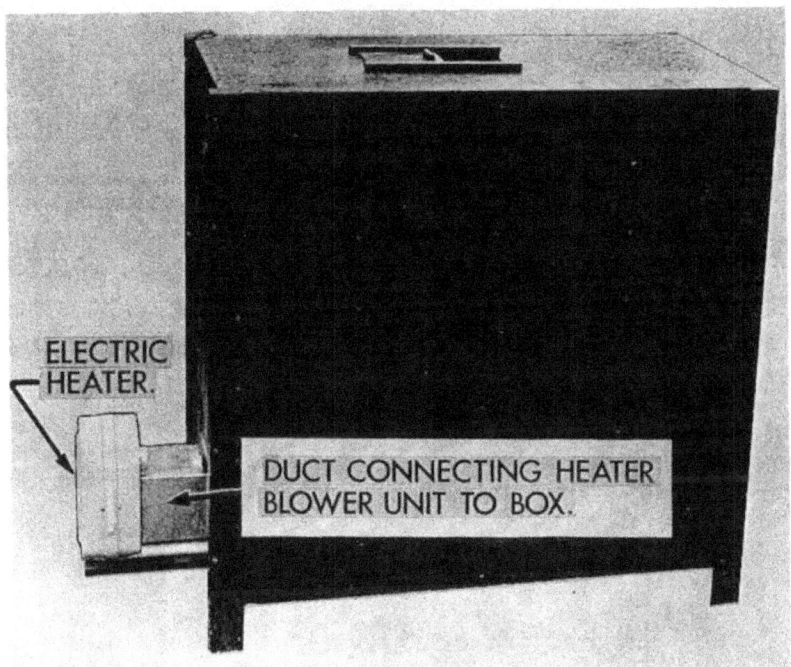

Figure 3.—A drying box, showing heating unit and duct; top shows sliding door used for air circulation.

Processing before drying.—Roots are usually sliced lengthwise or crosswise to hasten drying and to minimize spoilage and molding. (Ginseng roots are not sliced because their shape is important in meeting market demands.) Fleshy fruits, which are particularly apt to spoil, should be cut in halves or quarters and dried in a drying box. Bark can be cut into uniform pieces to hasten drying. Seeds should be spread in a fairly thin layer in aluminum or cardboard pie plates or similar available containers.

Figure 4.—Interior of an artificial drying box, showing drying racks.

Packaging and Storing

Clean burlap sacks, boxes, and paper sacks are all usable for packing dried plant material. Boxes should be dry and lined with clean paper. Collectors should avoid using plastic bags because any excess moisture present when the bags are shut may result in molding.

Plant material should be stored under sanitary conditions that minimize rodent and insect contamination. Clean, dry, ventilated storage areas are best for preserving quality, (fig. 5).

Figure 5.—A buyer of medicinal plants inspects a drying spikenard root.

COLLECTING POLLEN

Pollen is used by drug companies for making preparations to test for pollen allergies. Among the pollens most in demand are ragweeds, sages, magwort, sagebrush, elm, box elder, maple, ash, oak, cocklebur, pigweed, and Russian thistle.

Different kinds of pollen should not be mixed together because pollen buyers inspect shipments with a microscope, and any impure materials are rejected. Pollen should not be collected from plants that have been treated with pesticides.

Methods

The day before pollen is to be collected, tie several blooming heads together with white string to mark the plants wanted and to reduce loss from wind (fig. 6).

Pollen can be harvested from a plant for several days, but when the pollen begins to turn dark another plant should be used.

Harvesting can begin on clear mornings, as soon as the dew is gone; and, depending on the wind, harvesting can continue for about 2 hours. However, if the day is still, harvesting can continue longer.

Figure 6.—One way to collect pollen: roll the heads of the flowers gently over a catching cloth, spreading the blooms with your fingers to release the pollen.

Drying

Immediately after a day's harvesting is finished, the pollen should be spread out on clean, dry, brown wrapping paper (an opened grocery bag will do) in a warm, draft-free room. The pollen should be spread to a depth of about 1/4 inch and left to air-dry for 4 days. Mold may occur on the pollen if it is dried less than 4 days, and moldy pollen will not be accepted by buyers.

When dry, the pollen should be strained through fresh nylon or organdy, and packed in clean, dry, screw-top jars or in clean, dry, strong plastic bags.

Grass Pollen

Because pollen from grasses such as timothy, Johnson and others is difficult to collect in the field, a special harvesting technique has been worked out. As they mature heads of plants are harvested in the field and brought to a shed or protected areas. The stems are placed in a container of water and the pollen is collected on sheets of paper (fig. 7) placed next to the container. The pollen is then cleaned through nylon or organdy and packed for shipment.

Figure 7.—Grass pollen is collected on sheets of paper after mature heads of plants have been harvested.

REFERENCES

Bailey, L. H.
 1951. MANUAL OF CULTIVATED PLANTS. Revised ed., 1116 pp., illus. Macmillan Co., New York.

Burn, Harold.
 1962. DRUGS, MEDICINE AND MAN. 248 pp. Charles Scribner's Sons, New York.

Claus, Edward P., and Varro E. Tyler, Jr.
 1965. PHARMACOGNOSY. Ed. 5, 572 pp., illus. Lea & Febiger, Philadelphia.

Collingwood, G. H., and
Warren D. Brush.
 1937. KNOWING YOUR TREES. 109 pp. Amer. Forestry Assoc., Washington, D. C.

Coon, Nelson.
 1963. USING PLANTS FOR HEALING. 272 pp. Heartside Press, Inc., New York.

Curtin, L. S. M.
 1947. HEALING HERBS OF THE UPPER RIO GRANDE. 281 pp., illus. Lab. Anthrop., Santa Fe, N. M.

Darlington, William.
 1859. AMERICAN WEEDS AND USEFUL PLANTS. 460 pp., illus. A. O. Moore & Co., New York.

Edwards, Bertie.
 [n.d.] METHODS OF COLLECTING, DRYING, CLEANING, AND SELLING POLLEN. Bull. 1. Lenoir, N. C.

Fernald, Merritt Lyndon, and
Alfred Charles Kinsey.
 1943. EDIBLE WILD PLANTS OF EASTERN NORTH AMERICA. 452 pp., illus. Cornwall Press, New York.

Fernald, Merritt Lyndon.
 1950. GRAY'S MANUAL OF BOTANY. Ed. 8, 1632 pp., illus. American Book Co., New York.

Ford, Thomas R.
 1962. THE SOUTHERN APPALACHIAN REGION. 308 pp. Univ. Ky. Press, Lexington, Ky.

Fowells, H. A.
 1965. SILVICS OF FOREST TREES OF THE UNITED STATES. U. S. Dep. Agr. Agr. Handb. 271, 762 pp., illus.

Gibbons, Euell.
 1966. STALKING THE HEALTHFUL HERBS. 303 pp., illus. David McKay Co., New York.

Gleason, Henry A.
 1952. THE NEW BRITTON AND BROWN ILLUSTRATED FLORA OF THE NORTHEASTERN UNITED STATES AND ADJACENT CANADA. 1733 pp., illus. (3 vols.) N. Y. Botanical Garden, New York.

Gosselin, Raymond.
 1962. THE STATUS OF NATURAL PRODUCTS IN THE AMERICAN PHARMACEUTICAL MARKET. Lloydia 24(4):241-243.

Greer Drug and Chemical Corporation.
 [n.d.] INSTRUCTIONS FOR COLLECTING AND DRYING SHORT RAGWEED POLLEN. 13 pp. Lenoir, N. C.

Grieve, M.
 1959. A MODERN HERBAL. 888 pp., illus. (2 vols.) Hafner Publ. Co., New York.

Hardin, James W.
 1961. POISONOUS PLANTS OF NORTH CAROLINA. N. C. State Coll. Agr. Exp. Sta. Bull. 414, 128 pp.

Hardin, James W.
 1964. NORTH CAROLINA DRUG PLANTS OF COMMERCIAL VALUE. N. C. State Coll. Agr. Exp. Sta. Bull. 418, 34 pp.

Harding, A. R.
 1936. GINSENG AND OTHER MEDICINAL PLANTS. 367 pp., illus. A. R. Harding Publ., Ohio.

Hocking, George.
 1955. A DICTIONARY OF TERMS OF PHARMACOGNOSY AND OTHER DIVISIONS OF ECONOMIC BOTANY. 484 pp. Charles C. Thomas Publ., Bannerstone House, Ill.

Imbesi, A.
 1964. INDEX PLANTARUM QUET IN OMNIUM POPULORUM PHARMACOPOEIS SUNT. 771 pp. Adhuc Receptae, Messina, Sicily, Italy.

Jacobs, Marion Lee, and
Henry M. Burlage.
 1958. INDEX OF PLANTS OF NORTH CAROLINA WITH REPUTED MEDICINAL USES. 322 pp. Chapel Hill, N. C.

Jaques, H. E.
 1959. HOW TO KNOW THE WEEDS. 230 pp., illus. Wm. C. Brown Co., Dubuque, Iowa.

Kelsey, Harlen P., and
William A. Dayton.
 1942. STANDARDIZED PLANT NAMES. Ed. 2, 675 pp. J. Horace McFarland Co., Harrisburg, Pa.

Kingsbury, John M.
 1964. POISONOUS PLANTS OF THE UNITED STATES AND CANADA. 626 pp. Prentice-Hall, Inc., Englewood Cliffs, N. J.

Kreig, Margaret.
 1964. GREEN MEDICINE. 462 pp., illus. Rand McNally & Co., Chicago.

Krochmal, Arnold.
 1968. MEDICINAL PLANTS IN APPALACHIA. Econ. Bot. 22(4):332-337.

Little, Elbert L., Jr.
　1953. CHECK LIST OF NATIVE AND NATURALIZED TREES OF THE UNITED STATES (INCLUDING ALASKA). U. S. Dep. Agr. Agr. Handb. 41, 472 pp.

Massey, A. B.
　1942. MEDICINAL PLANTS. Va. Polytech. Inst. Bull. 30, 52 pp,. illus.

Meyer, James F.
　1960. THE HERBALIST. 304 pp., illus. Rand McNally & Co., New York.

Miller, James F.
　[n.d.] WEED IDENTIFICATION. 97 pp., illus. Univ. Ga. Coop. Ext. Serv.

Osol, Arthur, and George Farrar.
　1950. THE DISPENSATORY OF THE UNITED STATES OF AMERICA. Ed. 24, 2155 pp. (2 vols.) J. P. Lippincott Co., Philadelphia.

Osol, Arthur, Robertson Pratt, and Mark D. Altschule.
　1967. THE UNITED STATES DISPENSATORY. Ed. 26, 1277 pp. J. P. Lippincott Co., Philadelphia.

Quer, P. Font.
　1962. PLANTAS MEDICINALES. 1033 pp., illus. Editorial Labor, S. A., Barcelona, Spain.

Radford, A. W., H. E. Ahles, and C. R. Bell.
　1964. GUIDE TO THE VASCULAR FLORA OF THE CAROLINAS. 383 pp. Univ. N. C. Bot. Dept., Raleigh.

Sargent, C. R.
　1965. MANUAL OF THE TREES OF NORTH AMERICA. 1367 pp., illus. (2 vols.) Dover Publ., New York.

Shelton, Ferne.
　1965. PIONEER COMFORTS AND KITCHEN REMEDIES. 24 pp. Hutcraft, High Point, N. C.

Steinmetz, E. F.
　1957. CODES VEGETABILIS. 149 pp. Steinmetz Publ., Amsterdam, Netherlands.

Steinmetz, E. F.
　1959. DRUG GUIDE 1959. 382 pp. Steinmetz Publ., Amsterdam, Netherlands.

Strausbaugh, P. D., and Earl L. Core.
　1952. FLORA OF WEST VIRGINIA (Part I). W. Va. Univ. Bull. 52 (12):1-273, illus.

Strausbaugh, P. D., and Earl L. Core.
　1953. FLORA OF WEST VIRGINIA (Part II). W. Va. Univ. Bull. 53 (12):275-570, illus.

Strausbaugh, P. D., and Earl L. Core.
　1958. FLORA OF WEST VIRGINIA (Part III). W. Va. Univ. Bull. 58 (12):571-860, illus.

Strausbaugh, P. D., and Earl L. Core.
　1964. FLORA OF WEST VIRGINIA (Part IV). W. Va. Univ. Bull. 65 (3):861-1075, illus.

Strausbaugh, P. D., and Earl L. Core.
　1964. FLORA OF WEST VIRGINIA (Introductory Section). W. Va. Univ. Bull. 65 (3):i-xxxi.

Subcommittee on Standardization of Common and Botanical Names of Weeds.
　1966. WEEDS. 14 (4):347-386.

Tehon, Leo R.
　1951. THE DRUG PLANTS OF ILLINOIS. Ill. Nat. Hist., Surv. Cir. 44, 135 pp., illus.

Todd, R. G. (ed.).
　1967. EXTRA PHARMACOPEIA. 1804 pp. Pharmaceutical Press. London.

Williams, Louis O.
　1960. DRUG AND CONDIMENT PLANTS. U. S. Dep. Agr. Agr. Handb. 172, 37 pp., illus.

Useful Serials

Acta Phytotherapeutica.
　1954-68. E. F. Steinmetz, Publ., Amsterdam.

American Journal of Pharmacy.
　1825-68. Phila. Coll. Pharm. & Sci., Philadelphia.

American Perfume and Essential Oil Review.
　1906-68. New York.

Bibilography of Forest and Forestry Products.
　1948-68. Food and Agr. Organ., United Nations, Rome.

Biologia (monthly supl. to Chronica Botanica).
　1947-68. Waltham, Mass.

Bulletin of Miscellaneous Information.
　1896-1954. Roy. Bot. Gard., Kew, London.

Bulletin of the Lloyd Library and Museum of Botany, Pharmacy, and Materia Medica.
　1934-68. Cincinnati.

Chemurgic Digest.
　1942-68. Nat. Farm Chemurgic Council. New York.

Digest of Comments on the Pharmacopoeia of the United States and on the National Formulary.
　1905-22. U. S. Public Health Serv. Hygienic Lab. Bull., Washington, D.C.

Drug and Cosmetic Industry.
 1914-68. Pittsburgh, Mass., and New York.

Drug Topics.
 1883-1968. New York.

Drug Trade News.
 1925-168. New York.

Drug Treatises.
 1904-11. Lloyd Bros., Inc., Cincinnati.

Economic Botany.
 1947-68. Bot. Gard., New York.

Excerpta Botanica Sectio A.
 1959-68. Gustav Fischer, Stuttgart.

Farmacognosia.
 1938-68. Instituto Jose Celestino Mutis, de Farmacognosia, Madrid.

Fitotherapia.
 1929-68. Inverni & Della Beffa S.p.a., Milan.

Lloydia.
 1937-68. Lloyd Library, Cincinnati.

Qualitas Plantarum et Materiae Vegetabiles.
 1953-68. W. Junk, publ., The Hague.

Quarterly Journal of Crude Drug Research.
 1961-68. E. F. Steinmetz, publ., Amsterdam.

GLOSSARY

Botanical and Pharmacological Terms

Acute. Sharp-pointed.
Ague. Old word for fever, usually malaria.
Allergenic. Produces allergy.
Alterative. Changes a condition gradually.
Ament. Catkin.
Annual. A plant that completes its development from germination of the seed through flowering and death in one growing season.
Anodyne. Relieves or quiets pain.
Antacid. Neutralizes excess acidity in the alimentary canal.
Anthelmintic. Capable of expelling or destroying intestinal worms.
Antiasthmatic. Relaxes bronchial muscles and relieves labored breathing.
Antidiarrheal. Counteracts diarrhea.
Antidote. Counteracts the action of a poison.
Antiemetic. Lessens the tendency to vomit.
Anti-infective. Prevents or inhibits infection.
Anti-inflammatory. Reduces inflammation and swelling.
Antinauseant. Stops or lessens the tendency to become nauseated.
Antipruritic. Prevents or relieves itching (antipsoriatic).
Antipyretic. Reduces fever.
Antirheumatic. Reduces pain in the joints.
Antiseptic. Checks or inhibits the growth of microorganisms.
Antispasmodic. Reduces spasm or prevents convulsion.
Antitussive. Relieves or prevents coughing.
Aquatic. Growing in water.
Aromatic. Agreeable, usually spicy, odor.
Astringent. Causes the contraction of tissue.
Axil. Angle between stem and leaf stalk.
Axis. Main line of growth.
Balsam. An aromatic substance produced in certain plants.
Basal. Occurring at the bottom.
Biennial. A plant that requires 2 growing seasons to complete its development from germination of the seed through flowering and death.
Bract. Modified leaf, often below a flower petal.
Branchlet. A small branch growing from a large branch or tree trunk.
Bristly. Having short, stiff hairs.
Bur. Prickly seed envelope such as that of burdock (*Arctium*).
Buttressed. With projecting parts, usually refers to trunk of trees such as Cypress.
Capsule. A closed container bearing seeds; also a dry fruit with more than one part.
Carcinogenic. Causing cancer.
Cardio. Referring to heart action.
Carminative. Used to relieve gas and colic.
Catarrhal. Related to inflammation of the respiratory tract.
Cathartic. Causes an evacuation of the bowel.
Catkin. A scaly, drooping spike of flowers, such as that of willow.

Caustic. Destroys tissue.
Central nervous depressant. Depresses central nervous system activity.
Central nervous stimulant. Increases central nervous system activity.
Cholagogue. Increases the flow of bile.
Clasping. Partly or wholly surrounding the stem.
Cleft. With a space or division in the middle.
Clover-like. With leaves in three parts.
Cluster. A number of similar flowers or fruits growing closely together.
Composite. Refers to a structure apparently simple but made up of several distinct parts.
Compound. Two or more similar parts of a plant, especially fruits or leaves, united together into one whole.
Constituent. A component.
Corm. An enlarged solid bulb-like stem, usually underground.
Corolla. Usually petals.
Corona. An appendage borne between corolla and stamens in some flowers.
Corrective. Used to correct or make more pleasant the action of other remedies, especially purgatives. (Now called flavoring.)
Counterirritant. Causes irritation of the surface of an area with the object of relieving a deep-seated congestion.
Creeping. Spreading over the ground or other surface.
Cumarin. A toxic white crystalline lactone found in many plants; used to make perfume and soap.
Cylindrical. Having the form of a cylinder.
Cyme. Broad, flat flower cluster.
Cytotoxic. Poisonous to cells.
Demulcent. Substance used to protect or soothe the mucous membrane.
Dental obtundant. Used to dull or soothe acute toothache.
Depurative. Removes impurities and waste materials and purifies the blood.
Detachable. Removable.
Diaphoretic. Used to increase perspiration.
Digestant. Aids in the digestion of foods.
Disinfectant. Destroys or inhibits the growth of harmful microorganisms.
Diuretic. Increases the volume of urine.
Dormant. Resting or non-vegetative stage, usually during winter.
Downy. Covered with soft hairs.
Drab. Dull brown, or gray.
Drupe. Fleshy seeded fruit with one seed enclosed in a stony cover; peach, apricot.
Dyspepsia. A disturbed digestive condition characterized by nausea, gas, and heartburn.
Ellipsoid. Solid with elliptical outline.
Elliptical. Shaped like an elongated circle.
Elongate. Stretched out.
Emetic. An agent that causes vomiting.
Emmenagogue. An agent that induces menstrual flow.
Emmolient. Used externally to soften the skin and protect it.
Enzyme. Organic substance causing chemical changes without undergoing any chance of its own.
Excrescence. An outgrowth or enlargement.

Expectorant. An agent that causes expulsion of mucous from respiratory tract.
Exudate. Discharge in layers or flakes.
Febrifuge. Reduces fever (antipyretic).
Flatulence. Stomach discomfort caused by gas.
Frond. Leaf of fern or palm.
Fungicide. An agent that destroys fungi.
Furrowed. Wrinkled, corrugated, grooved.
Garbling. Process of sorting out and cleaning the usable parts of plants.
Genera. Groups of related plants.
Habitat. Particular location where plant usually grows.
Hemostatic. An agent used to stop internal hemmorrhage.
Herbaceous. Dying down annually at onset of winter.
Herb. Leafy upper portion of plant, minus roots.
Humus. Organic portion of the soil, usually dark colored.
Husk. Outer covering of seed or fruits.
Hypnotic. An agent that induces sleep without delirium.
Incision. A sharp, narrow notch or separation, as in the margin of a leaf.
Insecticide. An agent that kills insects.
Intoxicant. An agent that produces mental confusion with subsequent loss of muscular control.
Irritant. Causes inflammation of, or stimulation to, the tissues.
Lanceolate. Much longer than broad; lance-shaped.
Lateral. Occurring on a side.
Laxative. A cathartic that causes a more or less normal evacuation of the bowel without griping or irritation.
Leaflet. Part of a compound leaf.
Leafstalk. Stem of a leaf.
Linear. Going in a straight line.
Lobe. Rounded part or segment of an organ, usually part of a leaf or petal.
Mucilaginous. Slimy.
Narcotic. An agent that relieves distress and induces sleep.
Nodding. Drooping.
Node. The often swollen point on a stem at which a leaf is joined.
Oblanceolate. Having the broadest part of a lanceolate body above the middle.
Oblong. Longer than broad.
Opposite. Situated in pairs on an axis, each being separated from the other by half the circumference of the axis.
Opthalmiatric. Used in the treatment of eye diseases.
Ovate. Resembling hen eggs split lengthwise.
Palmate. Resembling a hand with fingers spread.
Panicle. Loosely branched flower cluster, pyramidal shaped.
Parasiticide. An agent that destroys animal or vegetable parasites.
Pectoral. Usually an expectorant, used for diseases of the chest and lungs.
Perennial. Continuing or lasting for several years.
Petal. Usually colored part of a flower.
Petiole. Leafstalk.
Pod. A dry seed vessel or fruit.
Pollen. Shed by male flowers, usually yellow dust; male reproductive agent.

Protective. Used locally to protect and soothe the skin and mucous membranes.
Pungent. A sharp sensation as to taste, smell, feeling.
Purgative. Increases peristalsis (contraction of the bowel).
Pustulant. Causes severe irritation of the skin, especially the sweat glands, and results in pustule formation.
Raceme. An elongated axis bearing flowers on short stalks.
Reclining. Bent down.
Refrigerant. Allays thirst and gives a sensation of coolness to the body.
Resinous. Characteristic of resin, a solid to semi-solid yellowish brown plant substance.
Respiratory sedative. Used to allay coughs.
Respiratory stimulant. Stimulates the respiratory centers.
Rhizome. Underground stem.
Rhombic. Having the form of an equilateral parallelogram.
Rootstock. Rhizome.
Rosette. Leaves orginating from a center point, or short intermode, often close to the ground.
Ross. To remove coarse outer bark.
Rubifacient. Causes reddening and mild irritation of the skin.
Saprophyte. A plant living on dead or decaying plant material.
Scale. Small leaves or bracts.
Sedative. Used to quiet the individual.
Serrate. Saw-toothed margin of a leaf.
Sessile. Lacking a stalk, hence directly attached to a main stem or branch.
Sheath. A long or tubular structure surrounding a stem.
Sialagogue. Causes an increase in the flow of saliva.
Simple leaf. A leaf that is not divided into leaflets even though lobed.
Solitary. Borne alone.
Somnifacient. Produces sleep without delirium; a soporific.
Soporific. Tending to cause sleep.
Spike. Usually an axis bearing flowers without stalks.
Spiny. Bearing sharp-pointed prickles or woody bodies.
Stalk. Stem on which a leaf, flower, or other organ is attached.
Stimulant (cerebral). An agent that stimulates the activity of the cerebellum, especially the centers of reason, thought, etc.
Stimulant (general). A substance which increases general functional activity.
Stomachic. Stimulates appetite and increases secretion of digestive juices.
Strict. Straight and upright; few if any branches.
Subtend. Below and close to, such as a bract below a petal.
Sudorific. Increases perspiration.
Taeniafuge. A tapeworm expellant.
Taenicide. As agent that destroys tapeworms.
Taproot. A main root growing down, with small lateral roots.
Terminal. At the tip.
Tonic. Stimulates the restoration of tone to the muscles.
Toothed. Indented.
Trifoliate. Having three leaflets.
Tubular. Tube-shaped; hollow cylinder.
Tufted. Having small bunches of hair close together.
Twining. Twisting and winding.

Umbel. A flat-topped cluster of flowers arising from a common point.
Urinary antiseptic. Retards the growth of microorganisms in the urinary tract.
Vasoconstrictor. Narrows the passageway of the blood vessel.
Vermicide. An agent that destroys worms.
Vesicant. Causes irritation to the skin, resulting in blisters.
Viscous. Sticky and thick.
Vulnerary. An agent that promotes healing of open wounds.
Whorl. Three or more flowers or leaves at a node forming a circle.
Winged. Having wings, such as the thin dry extensions on a maple seed.

Meanings of Terms Used in Plant Names

Acutiloba. Having sharp lobes.
Alba. White.
Albidum. Whitish.
Ambrosioides. Fragrant, like ambrosia.
Americanus. American.
Androsaemifolium. Having leaves like those of Androsaemum.
Aparine. Bedstraw.
Arborescens. Tree-like.
Atropurpurea. Very dark purple.
Aureus. Gold.
Balsamifera. Producing balsam.
Benedictus. Blessed.
Benzoin. A plant of the laurel family.
Biflorum. Having two flowers.
Calamus. Reed.
Canadense. Of Canada.
Capillus veneris. Hair-like.
Cardiaca. Heart-like.
Carolinense. Of Carolina.
Cataria. Catnip.
Cerifera. Wax-producing.
Cinerea. Grayish.
Clava-Herculis. Hercules club.
Communis. In groups.
Cordifolia. Heart-shaped.
Crispus. Waved and twisted.
Didyma. In pairs.
Diphylla. Two-leaved.
Erectum. Erect.
Farinosa. Covered with whitish mealy powder.
Frondosa. Full of leaves.
Glabra. Smooth.
Hastata. Triangular halberd-shaped lobes.
Hippocastanum. Horse-chestnut.
Hybridus. Mixed or impure.
Hydropiper. Water pepper.
Hyemale. Of the winter evergreen.

Incarnata. Flesh-colored.
Inflata. Expanded.
Lappa. Bur-like.
Lateriflora. Having flowers on the side.
Lenta. Pliant, tough.
Luteum. Yellow.
Maculata. Spotted.
Marilandica. Of Maryland.
Medica. Middle.
Millefolium. Very many leaved.
Minus. Lesser or smaller.
Nigra. Black.
Nudicaulis. Naked-stemmed.
Nudum. Bare.
Odoratissima. Very fragrant.
Officinale. Used medically.
Palustris. Of swamps.
Parviflorum. Small-flowered.
Pedatum. Like a bird's foot.
Peltatum. Shield-shaped.
Peregrina. Traveling from a strange country.
Perfoliatum. Having pierced leaves.
Piperita. Peppery.
Pratense. Of meadows.
Procumbens. Flat, prostrate.
Prunifolium. Plum-like leaves.
Pulegioides. Like Pennyroyal.
Quinquefolium. Five-leaved.
Racemosa. Full of clusters.
Repens. Creeping.
Rubra. Red.
Scariola. Papery, scaly.
Sempervirens. Evergreen.
Serotina. Late-flowering.
Serpentaria. Snake bite cure.
Serrulata. Finely serrated.
Simplicissima. Undivided.
Spicata. Bearing a spike.
Stramonium. Swelling.
Strobus. Overlapping scales; cone.
Sylvatica. Of the forest, wild.
Styraciflua. Flowering gum.
Syriaca. Of Syria.
Thalictroides. Like meadow rue.
Thapsus. Of ancient Thapsus.
Tinctoria. Of dyes.
Triphyllum. Three-leaved.
Tuberosa. Having tubers.
Umbellata. Having flowers arranged in umbels.
Villosa. Shaggy, hairy.
Viride. Green.
Vulgare. Common.

www.ingramcontent.com/pod-product-compliance
Lightning Source LLC
Chambersburg PA
CBHW052238150526
45153CB00029B/295